PROTECTING NUCLEAR WEAPONS MATERIAL
IN RUSSIA

Office of International Affairs

National Research Council

NATIONAL ACADEMY PRESS

Washington, D.C.

NATIONAL ACADEMY PRESS · 2101 Constitution Ave., N.W. · Washington, D.C. 20418

NOTICE: The project that is the subject of this report was approved by the Governing Board of the National Research Council, whose members are drawn from the councils of the National Academy of Sciences, the National Academy of Engineering, and the Institute of Medicine. The members of the committee responsible for the report were chosen for their special competencies and with regard for appropriate balance.

The National Academy of Sciences is a private, nonprofit, self-perpetuating society of distinguished scholars engaged in scientific and engineering research, dedicated to the furtherance of science and technology and to their use for the general welfare. Upon the authority of the charter granted to it by Congress in 1863, the Academy has a mandate that requires it to advise the federal government on scientific and technical matters. Dr. Bruce M. Alberts is president of the National Academy of Sciences.

The National Academy of Engineering was established in 1964, under the charter of the National Academy of Sciences, as a parallel organization of outstanding engineers. It is autonomous in its administration and in the selection of its members, sharing with the National Academy of Sciences the responsibility for advising the federal government. The National Academy of Engineering also sponsors engineering programs aimed at meeting national needs, encourages education and research, and recognizes the superior achievements of engineers. Dr. William Wulf is president of the National Academy of Engineering.

The Institute of Medicine was established in 1970 by the National Academy of Sciences to secure the services of eminent members of appropriate professions in the examination of policy matters pertaining to the health of the public. The Institute acts under the responsibility given to the National Academy of Sciences by its congressional charter to be an adviser to the federal government and, upon its own initiative, to identify issues of medical care, research, and education. Dr. Kenneth I. Shine is president of the Institute of Medicine.

The National Research Council was organized by the National Academy of Sciences in 1916 to associate the broad community of science and technology with the Academy's purposes of furthering knowledge and advising the federal government. Functioning in accordance with general policies determined by the Academy, the Council has become the principal operating agency of both the National Academy of Sciences and the National Academy of Engineering in providing services to the government, the public, and the scientific and engineering communities. The Council is administered jointly by both Academies and the Institute of Medicine. Dr. Bruce M. Alberts and Dr. William Wulf are chairman and vice-chairman, respectively, of the National Research Council.

This project was sponsored by Brookhaven National Laboratory. Any opinions, findings, conclusions, or recommendations expressed in this publication are those of the authors and do not necessarily reflect the view of the organization or agencies that provided support for the project.

International Standard Book Number 0-309-06547-X

A limited number of copies of this report are available from:

Division on Development, Security, and Cooperation
National Research Council
2101 Constitution Avenue, N.W. FO2060
Washington, D.C. 20418
Tel: (202) 334-2644

Copies of this report are available for sale from:

National Academy Press
2101 Constitution Avenue, N.W. Box 285
Washington, D.C. 20055
Tel: 1-800-624-6242 or (202) 334-3313 (in the Washington Metropolitan Area).

CONTENTS

PREFACE

ORIGIN OF THE STUDY

In April 1997, the National Research Council (NRC) published *Proliferation Concerns: Assessing U.S. Efforts to Help Contain Nuclear and Other Dangerous Materials and Technologies in the Former Soviet Union*. One of the two sections of the report provided an assessment of the significance and effectiveness of cooperative programs of the Department of Energy (DOE) to upgrade material protection, control, and accountability (MPC&A) for "direct-use" material (defined to include unirradiated highly enriched uranium and separated plutonium) in Russia, Kazakhstan, Ukraine, and Belarus. The relevant recommendations of that report are set forth in Appendix A.

In the spring of 1998, DOE requested an updated assessment of its MPC&A activities in Russia (see Appendix B for the Terms of Reference of the Study). DOE believed that the possibility of theft or diversion of direct-use material was more serious than estimated several years ago. Through on-the-ground experience, DOE had learned that direct-use material was dispersed in many more locations than previously estimated and that upgrading MPC&A capabilities was much more complicated than anticipated. Moreover, the expected improvement or at least stabilization of the Russian economy had not occurred; in fact, beginning in August 1998, there was a precipitous decline in economic conditions. DOE requested a focus on Russia because most of the material is located in that country, including much material that has not yet been brought into the DOE program; this situation is in contrast to DOE activities in the other three countries, which encompass all known direct-use material. In response to the request, the assessment began in September 1998, and the findings and recommendations are set forth in this report.

In carrying out its work, the committee reviewed recent DOE activities related to the recommendations in *Proliferation Concerns*. In many cases, DOE had taken steps consistent with the recommendations, although implementation of these recommendations has been uneven. In any event, the committee has reiterated those recommendations that remain important and has provided up-to-date justifications for their implementation. Although building on the earlier report, this report is intended as a stand-alone assessment as of March 1, 1999.

RELATED STUDIES BY OTHER ORGANIZATIONS

This study is a complement to an internal review of DOE's MPC&A activities in Russia being carried out by Brookhaven National Laboratory (BNL). Although the report of the internal review has not yet been published, there is considerable consistency between the preliminary recommendations released by the Brookhaven team and the recommendations in this report. However, this report is not intended to be a critique of the BNL recommendations, but rather it is an independent review of many of the same issues highlighted by the BNL study.[1]

A number of nongovernmental organizations have an interest in this topic. Among the most active organizations are Harvard University, the Monterey Institute of International Studies, Princeton University, the Russian–American Nuclear Security Advisory Committee, and the Union of Concerned Scientists. The views of individuals involved in their activities have been taken into account and have been quite helpful in preparing this report.

ROLE OF THE NRC COMMITTEE FOR THIS STUDY

In September 1998, the Chairman of the NRC appointed a four-person interdisciplinary committee to carry out this study. The members had served on the committee responsible for the earlier study and are identified in Appendix C.

During the fall of 1998, committee members traveled to Moscow, the Moscow region, and Dmitrovgrad to observe completed projects and work in progress. Committee members also visited Lawrence Livermore, Los Alamos, Oak Ridge, and Sandia National Laboratories where many of the U.S. specialists who participate in the program are based. Representatives of DOE, BNL, and the Russian–American Nuclear Security Advisory Committee briefed the committee.

Throughout the entire process, many officials and specialists in the United States and Russia took time to provide important information and insights to the committee. DOE was extraordinarily helpful in arranging visits and consultations. Appendix D identifies the formal meetings and visits. Of no less significance were the many informal discussions also arranged through numerous channels in the United States and abroad. Finally, DOE and other organizations provided the committee with a wealth of documents, and the most significant ones are listed in the Bibliography.

---

[1] For the preliminary recommendations by BNL, see C. Ruth Kempf, "Russian Nuclear Material Protection, Control, and Accounting Program: Analysis and Prospect," *Partnership for Nuclear Security* (Washington, D.C.: Office of Nonproliferation and Arms Control, September 1998), pp. 40-44.

This report has been reviewed in draft form by individuals chosen for their diverse perspectives and technical expertise, in accordance with procedures approved by the NRC's Report Review Committee. The purpose of such an independent review is to provide candid and critical comments that will assist the institution in making the published report as sound as possible and to ensure that the report meets institutional standards for objectivity, evidence, and responsiveness to the study charge. The review comments and draft manuscript remain confidential to protect the integrity of the deliberative process. I wish to thank the following individuals for their participation in the review of this report: Harold Agnew (General Atomics, ret.), Gary Bertsch (University of Georgia), Matthew Bunn (Harvard University), Harold Forsen (Bechtel Corporation, ret.), William Hannum (Argonne National Laboratory), Kaye Lathrop (Stanford University, ret.), Albert Narath (Lockheed Martin, ret.), and Frank Parker (Vanderbilt University). These individuals have provided constructive comments and suggestions, but it must be emphasized that responsibility for the final content of this report rests entirely with the authoring committee and the institution.

Finally, the committee expresses its appreciation to the many individuals and institutions in the United States and Russia that assisted its efforts. It also is grateful for the exceptional assistance of the NRC staff.

Richard A. Meserve, Chairman, Committee on Protection, Control, and Accountability of Nuclear Materials in Russia

EXECUTIVE SUMMARY

A major technical impediment confronting a nation or group bent on developing nuclear weapons is the difficulty of obtaining the necessary direct-use material. A minimum of a few kilograms of plutonium or several times that amount of highly enriched uranium (HEU) is required, with the quantity depending on the composition of the material, type of weapon, and sophistication of the design.

Russia is estimated to have approximately 675 metric tons of such material outside nuclear weapons (75 metric tons of plutonium and 600 metric tons of HEU) at a variety of institutions, much of which is protected by only limited security measures. It is in the national security interest of the United States to join with Russia to strengthen the protection of this material.

A 1997 report of the National Research Council (NRC) entitled *Proliferation Concerns: Assessing U.S. Efforts to Help Contain Nuclear and Other Dangerous Materials and Technologies in the Former Soviet Union* discussed the inadequate protection of direct-use material in Russia at that time and the importance of U.S. efforts to help secure this material. Although progress has been made in improving the security arrangements at some sites during the past several years, the gravity of the threat has increased. The recent decline in the Russian economy has severely affected the economic well-being of many Russian government officials, nuclear specialists, and workers who have access or could arrange access to direct-use material. While direct-use material must be guarded closely even in the best of economic times, the level of economic deprivation has increased the likelihood of attempted thefts or diversions of such material from Russian facilities. Furthermore, expanded access to Russian facilities by U.S. specialists has provided the U.S. government with new insights into the vast Russian nuclear complex. The U.S. government has identified more extensive dispersion of material and more pervasive inadequacies of protection systems than had been anticipated. Thus, with the latest economic crisis and greater problems in ensuring the security of direct-use material, the threats of theft or diversion are considerably greater than estimated three years ago.

During the past several years, the Department of Energy (DOE) has carried out a program of cooperation with the Russian Ministry of Atomic Energy (MINATOM and other Russian organizations in the protection, control, and accountability of direct-use material (MPC&A). DOE has budgeted $140 million for the program during fiscal year 1999; and DOE has requested an appropriation of $145 million for the MPC&A program during fiscal year 2000. The program has made significant contributions to upgrading security of direct-use material at a number of Russian locations and has stimulated the gradual development of a cadre of Russian specialists who are qualified to take responsibility for installing and operating MPC&A systems.

Several dozen buildings now are well equipped with security systems, and dozens more are currently being upgraded. Rapid strides have been made in developing a comprehensive program to protect the many tons of material produced for use in nuclear-powered submarines. At the Luch Production Organization outside Moscow, hundreds of kilograms of direct-use material that had been located in dozens of buildings have been consolidated into six locations. More than 30 railcars that transport direct-use material across long stretches of Russia are being upgraded to ensure proper protection. These are but a sampling of many important achievements of the MPC&A program, and they were only possible with the support of DOE. However, they are but a small beginning; adequate MPC&A systems have yet to be designed and then installed for protection of hundreds of tons of direct-use material dispersed in hundreds of buildings.

U.S. programs also have improved the skills of many specialists responsible for the operation of modern MPC&A systems at several dozen facilities. Formal training programs, as well as important on-the-job training activities, are increasing the size of the pool every month. Still, a much larger stable of qualified Russian specialists is needed to operate the MPC&A systems that are being installed, let alone systems that should be developed in the future.

In short, despite the progress, there is much that remains to be done. Given the increased threats to direct-use material in Russia, the demonstrated capability of the DOE programs to reduce the vulnerability of this material, and the improved understanding of the time and costs associated with installing MPC&A systems, **continued DOE involvement in strengthening MPC&A systems in Russia should be a high-priority national security imperative for the United States for at least a decade.** Meanwhile, the U.S. government must continue to emphasize the importance of MPC&A as a nonproliferation imperative at the highest political levels in Russia to achieve the final goal of ensuring that MPC&A systems are in place and operating effectively at all locations and are financially supported by the Russian government.

There are problems in need of immediate attention. The 1998 economic crisis in Russia has severely affected the Ministry of Interior (MVD) guard forces assigned to Russian facilities where direct-use material is located. At a number of facilities, the guards have encountered months of delay in receiving paychecks, have not had winter clothing for outside patrols, and have not had access to adequate meals. This is a serious concern because the physical protection systems are useless if guard forces are unavailable to respond to intrusions. Emergency measures by DOE to address these problems during the winter of 1998–1999, undertaken at a cost of about $600,000, are a necessary start in ensuring that the guards perform at a professional level despite economic hardships.

The economic turmoil in Russia also has affected the institutions with responsibility for installing and maintaining MPC&A systems. Some Russian institutes do not have the funds to pay salaries or to ensure the continuous functioning of power and communications systems needed for operation of modern detection, alarm, and related security devices. Until economic conditions improve, they will not be able to operate the systems as intended without some U.S. financial support. To date, DOE support largely has been limited to installation, but not operation, of MPC&A systems.

For the long term, indigenization of MPC&A activities is essential. As noted earlier, the U.S. program must end eventually, and the perpetuation of the systems must be a Russian responsibility. Therefore, it is imperative to nurture Russian "ownership" of the technical approaches that are pursued, to encourage increased reliance on Russian specialists to lead MPC&A efforts, and to develop improved Russian capabilities to provide MPC&A equipment and services. Limited steps toward indigenization have been taken by DOE, but the program remains under the heavy influence of U.S. specialists accustomed to U.S. approaches.

DOE has made substantial progress in initiating programs at many sites, including some of the most sensitive sites in Russia. Still, the program has been delayed by administrative problems encountered in Russia at the national and facility levels, such as (1) uncertainties as to the commitment of some Russian institutions to the program, (2) difficulties in gaining routine access for U.S. specialists to sensitive facilities, (3) lack of satisfactory procedures for ensuring recognition by Russian authorities of exemptions from tax and customs payments, (4) confusion as to Russian certification requirements for equipment that is to be used, and (5) Russian indecision concerning a national materials accountancy system. Also, on-the-ground technical problems arise that sometimes result in inappropriate approaches.

Although DOE's priorities are generally consistent with the most urgent needs in protecting direct-use material, several areas require attention by DOE.

- The most glaring deficiency is the lack of progress in installing and putting into operation material accountancy systems at Russian sites—including even the basic step of ensuring a complete and accurate inventory. Without such a system, there may be no way to detect whether material has been lost. While years will be necessary to complete this task, a more aggressive approach is warranted.

- With several important exceptions, only limited progress has been made in efforts to consolidate direct-use material into a fewer number of buildings, and almost no progress has been made in encouraging Russian facilities with little need for direct-use material to transfer excess supplies to other facilities. Consolidation offers the opportunity to strengthen MPC&A systems at lower costs.

- Neither DOE nor Russian institutions have developed strategies to ensure the long-term sustainability of the MPC&A systems. In particular, not enough attention has been focused on ensuring that adequate Russian resources will be devoted to maintaining the MPC&A system after the completion of the DOE program.
- There has been insufficient progress in providing transport systems and trucks that will ensure that direct-use material is secure during shipments within and between sites.

The management challenge in orchestrating a multitude of DOE headquarters, laboratory, and contractor personnel at about 50 sites in Russia is daunting. Steps are needed to maximize the return on U.S. expenditures, to reduce redundancy while ensuring adequate oversight, and to provide additional work incentives that will attract highly qualified specialists from the United States and Russia to participate in the program.

This report contains many recommendation to address these and related issues. The most important recommendations include:

Sustain the U.S. commitment to the program.

- *Maintain the current level of U.S. support ($145 million per year) for at least the next five years and be prepared to increase funding should particularly important opportunities arise. In addition, plan to continue an appropriately scaled program of cooperation thereafter, with the scope and duration of the program depending on both progress in installing MPC&A upgrades and economic conditions in Russia.*
- *Provide support for operational costs of selected aspects of the personnel and technical infrastructure at Russian institutes to help ensure that MPC&A systems that have been installed are operated and maintained as intended.*

Reassess priorities to address important vulnerabilities.

- *Review the languishing materials accountancy programs at all sites and, as part of adjusting overall program priorities, devote additional resources to improve and speed up performance in this area.*
- *Continue to consolidate storage areas for direct-use material whenever possible and give greater attention to the establishment of well-designed central storage facilities that serve more than one site.*
- *Expand the transportation program to provide a larger number of more secure vehicles to a variety of facilities, while ensuring the soundness of the procedures for tracking the movement of direct-use material.*

Indigenize MPC&A capabilities.

- *Increase the percentage of available U.S. funding that is directed to financing activities of Russian organizations with a concomitant declining percentage directed to supporting U.S. participants in the program. This could be accomplished by using Russian specialists from institutions with well-developed MPC&A capabilities to replace some U.S. members of teams supporting activities at Russian institutions with less-developed capabilities.*

- *Expand efforts to utilize Russian equipment and services whenever possible and to encourage Russian enterprises and institutes to increase their capabilities to provide high-quality equipment and associated warranties and services.*

Reduce impediments to effective cooperation.

- *Develop an improved political/legal framework for U.S.-funded MPC&A activities in Russia that ensures long-term stability for the program and exemptions from taxes, customs charges, and related fees.*

- *Establish in Moscow a DOE-MPC&A office that can troubleshoot and help overcome barriers to rapid progress and that can facilitate the coordination of MPC&A activities with other DOE programs.*

Improve management of U.S. personnel and financial resources.

- *Develop a clearer division of responsibility between DOE headquarters staff and specialists of the DOE laboratories. The division should recognize the lead role of headquarters in intergovernmental negotiations, formulation of general policy guidance, determination of priorities among sites, and financial oversight. It should recognize the role of the laboratories in providing advice to headquarters on the policy aspects of the program, in making technical decisions in accordance with headquarters' policy guidance and budgetary allocations, and in providing specialists who are responsible for the development and implementation of MPC&A upgrades.*

- *Coordinate MPC&A program activities with activities of related DOE programs to take advantage of opportunities for programs to reinforce one another.*

Despite many program accomplishments to date, the remaining MPC&A task is huge. Reducing the risk of illicit transfers of direct-use material to an acceptable level will take many years of steady effort. DOE is in a unique position to accelerate the effort and should be provided with the means to do so.

1

IMPORTANCE OF MODERN MPC&A SYSTEMS IN RUSSIA

AVAILABILITY OF DIRECT-USE MATERIAL

A major technical barrier confronting any nation or group seeking to develop a nuclear weapon is the acquisition of direct-use material.[2] Such material includes separated plutonium (plutonium) and unirradiated highly enriched uranium (HEU). This material can be used directly in weapons without the need for complicated chemical processing. Several kilograms of plutonium or several times that amount of HEU are the minimum required to construct a nuclear weapon, with the quantity depending on the composition of the material, the type of weapon, and the sophistication of the design. The U.S. government estimates that the current inventory of direct-use material in the Russia is about 150 metric tons of plutonium and 1,200 metric tons of HEU. About one-half of each of these quantities (i.e., 75 metric tons of plutonium and 600 metric tons of HEU) is incorporated into weapons and the other half is in various forms—particularly metals, oxides, solutions, and scrap—at many enterprises and institutes throughout Russia.[3]

This study does not address plutonium and HEU in weapons because the control of weapons raises issues that are distinct from those surrounding the security of direct-use material. Therefore, this study considers material in the custody of the Ministry of Atomic Energy (MINATOM), as well as within the

---

[2] The committee recognizes the difficulty of many countries in developing delivery systems as well as the vital role of a number of international regimes in limiting access to the technologies necessary for these systems.

[3] MPC&A Program Strategic Plan, Office of Arms Control and Nonproliferation, U.S. Department of Energy, January 1998, p. 2. While these figures are commonly cited as the amounts in the former Soviet Union, almost all of this material is in Russia. For more information on estimates of Russian stocks of HEU and Pu, see David Albright, Frans Berkout, and William Walker, Plutonium and Highly Enriched Uranium: World Inventories, Capabilities, and Policies (Oxford: Oxford University Press, 1997), pp. 50-59 and pp. 94-116.

authority of several other organizations, but not material in the custody of the Ministry of Defense, other than fuel for nuclear reactors of the navy.[4] This organizational boundary for the study is consistent with the proscribed scope of the effort because the Ministry of Defense usually has custody of weapons but, with the exception of naval fuel, usually does not have custody of other direct-use material.[5]

As the result of on-the-ground experience during the past several years, the Department of Energy (DOE) has gained new insights as to the vastness of the dispersion of direct-use material throughout the Russian nuclear complex and the inadequacies of material protection, control, and accountability (MPC&A) systems at many facilities. DOE considers the security deficiencies much greater, both in terms of the number of buildings that require upgrades and the extent of upgrades that are needed, than previously estimated. Current estimates are that over 400 buildings require enhanced MPC&A systems.[6] As indicated in Table 1-1, DOE categorizes the sites at which direct-use material is located, as follows:

- Defense-related sites: uranium and plutonium cities, the nuclear weapons complex, locations of maritime fuel.
- Civilian-related sites: large fuel facilities, reactor-type facilities.

During the Soviet era, the security over almost all direct-use material was very tight. The discipline and loyalty of managers, workers, and guards in the Soviet nuclear weapons complex were seldom in question. They were well paid and well respected within Soviet society, and, like all Soviet citizens, they were subject to surveillance by the KGB and other security agencies. Physical protection was based more on deployment of manpower than on use of technical devices; there were extensive guard forces to control travel across closed borders, into and out of closed cities, and into and out of closed facilities. The civilian portion of the nuclear complex also was under special security arrangements, although not as exacting as security in the military portion.[7]

The Soviets maintained primitive accounting systems for direct-use material at each facility, relying primarily on handwritten documentation and only occasionally on computer-based records. The documentation was not always

---

[4] The Department of Defense (DOD) has a separate program with the Ministry of Defense on improving security and accounting for nuclear warheads. This program is not reviewed in this report.

[5] The committee observes that there are other materials, some of which are outside the control of both MINATOM and MOD, that could be used in weapons with a limited amount of chemical processing. In particular, HEU in spent fuel rods that has low burnup rates and/or has been in storage for many years also may be an immediate proliferation threat.

[6] DOE briefing of committee staff, March 1999.

[7] For more information about Soviet-era security, see Oleg Bukharin, "Security of Fissile Materials in Russia," *Annual Review of Energy and the Environment*, Vol. 21, pp. 467-496, 1996.

TABLE 1-1 Sites of MPC&A Cooperation as of January 1998

DEFENSE RELATED SITES

Uranium and Plutonium Cities
1. Chelyabinsk-65/Ozersk, Mayak Production Facility
2. Tomsk-7/Seversk, Siberian Chemical Combine
3. Krasnoyarsk-26/Zheleznogorsk, Mining and Chemical Combine
4. Krasnoyarsk-45/Zelenogorsk, Uranium Isotope Separation Plant
5. Sverdlovsk-44/Novouralsk, Urals Electrochemical Integrated Plant

Nuclear Weapons Complex
6. Arzamas-16/Sarov, All-Russian Scientific Research Institute of Experimental Physics
7. Chelyabinsk-70/Snezhinsk, All-Russian Scientific Research Institute of Technical Physics
8. Avangard Plant
9. Sverdlovsk-54/Lesnoy
10. Penza-19/Zarechny
11. Zlatoust-36/Trekhgorny

Maritime Fuel
12. Navy Site 49
13. Navy 2nd Site Northern Fleet Storage Site
14. Navy Site 34
15. PM-63 Refueling Ship
16. PM-12 Refueling Ship
17. PM-74 Refueling Ship
18. Sevmash Shipyard
19. Icebreaker Fleet
20. Kurchatov Institute, Navy Regulatory Project, Navy Training Project

CIVILIAN SITES

Large Fuel Facilities
21. Elekstrostal Production Association Machine Building Plant
22. Novosibirsk Chemical Concentrates Plant
23. Podolsk, Scientific Production Association Luch
24. Dmitrovgrad, Scientific Research Institute of Atomic Reactors
25. Obninsk, Institute of Physics and Power Engineering
26. Bochvar All-Russian Scientific Research Institute of Inorganic Materials

Reactor-Type Facilities
27. Dubna, Joint Institute of Nuclear Research
28. Scientific Research and Design Institute of Power Technology
29. Moscow Institute of Theoretical and Experimental Physics
30. Moscow State Engineering Physics Institute
31. Karpov Institute of Physical Chemistry
32. Beloyarsk Nuclear Power Plant
33. Sverdlovsk Branch of Scientific Research and Design Institute of Power Technology
34. Khlopin Radium Institute
35. Tomsk Polytechnical University
36. Petersburg Nuclear Physics Institute
37. Krylov Shipbuilding Institute
38. Lytkarino Research Institute of Scientific Instruments
39. Norilsk
40. Baltisky Zavod

Source: U.S. Department of Energy

complete or easily retrievable, and there have been reports that some facilities kept material in reserve—off the books—to ensure that quotas for producing materials could be met. Moreover, the committee was informed that there were significant errors in the records (e.g., in one case, supplies of HEU were recorded as low-enriched uranium). In 1997 the U.S. intelligence community stated that "the Russians may not know where all their material is located."[8]

The end of the cold war and the prospect of significant nuclear arms reductions reduced the defense roles for many nuclear facilities, and the future of large segments of the Russian nuclear complex became uncertain. In the early 1990s, MINATOM began instructing its institutes to become self-supporting and less reliant on government funds. Ministry and institute budgets declined precipitously, many buildings deteriorated badly, and a number of laboratories simply closed. MPC&A activities at many institutes suffered very directly as reliable guards and other key security personnel with uncertain paychecks were recruited by private security firms, and the support of activities that did not generate income, such as MPC&A, was given low priority.

At the beginning of 1998, DOE highlighted a number of MPC&A deficiencies that were attributed both to the lingering Soviet legacy and to economic difficulties:

- lack of unified physical protection standards and inadequate defenses of buildings and facilities within site-perimeter fences;
- lack of portal monitors to detect fissile materials leaving or entering a site;
- inadequate central alarm stations and inadequate alarm assessment and display capabilities;
- inadequate protection of guards from small-arms fire and inadequate guard force communications;
- lack of material accounting procedures that can detect and localize nuclear material losses;
- inadequate measurements of waste, scrap, and hold-up nuclear materials during processing and inadequate accounting of transfers of nuclear materials between facilities;
- antiquated tamper-indicating devices (seals) on nuclear material containers that cannot guarantee timely detection of nuclear material diversion.[9]

As recounted in the 1997 report by the National Research Council, Russian officials have reported two dozen incidents of thefts and attempted thefts of nuclear-related items at their facilities, with the last ones occurring in 1994. All of the cases involved much smaller quantities of material than would be

---

[8] John Deutch, "The Threat of Nuclear Diversion," testimony to the Permanent Subcommittee on Investigations of the Senate Committee on Government Affairs, March 20, 1996.
[9] *MPC&A Program Strategic Plan*, p. 3.

necessary to make a nuclear weapon.[10] U.S. officials recently confirmed that there had been seven smuggling incidents during the early 1990s involving small amounts of weapons-usable material, which they suspected were stolen from sites in Russia and other countries of the former Soviet Union.[11] There have not been confirmed reports available to the committee of additional cases of theft or attempted theft. In light of the inadequacies of the existing MPC&A systems, however, the possibility of undetected thefts cannot be ruled out.

RUSSIAN ECONOMIC CRISIS OF 1998

During the summer of 1998, Russia approached the brink of economic collapse with the bankruptcy of leading Russian banks, defaults on foreign debts, devaluation of the ruble, and dramatic increases in the rate of inflation. Both Russian and Western investors took steps to withdraw considerable amounts of money from the economy, and foreign assistance agencies—under the leadership of the International Monetary Fund—reconsidered the viability of their lending and grant programs in Russia.

For individual Russians, this economic chaos resulted almost immediately in extended delays in receiving paychecks, a significant decline in purchasing power of paychecks when received, and losses and devaluations of personal savings. Already inadequate medical services deteriorated still further, and the reliability of heating and electrical systems declined. This latest round of economic problems resulted in termination of employment for tens of thousands of Russian workers, with more layoffs promised in the months ahead. MINATOM announced plans to downsize its nuclear complex, and the likelihood of job opportunities in the private sector dwindled.

Institutes and enterprises that possessed direct-use material were faced with many new financial problems. MINATOM and other government ministries that provided financial resources for the institutes saw their budgets severely slashed, and some foreign sources of financing hesitated to commit additional funds to Russia until the economic situation stabilized. Strikes erupted in Snezhinsk and other nuclear cities where paychecks were delayed. Some Russian institutes also did not have the funds to ensure the continuous functioning of power and communications systems needed for operation of modern detection, alarm, and related security devices.

The guard forces of the Ministry of Interior at facilities where direct-use material is located were particularly hard hit with the onset of winter. Some had no winter uniforms for outside patrols, and the heat in buildings often was turned off. Many were without paychecks, and they were no longer served adequate meals as the budgets for support disappeared. The committee heard

---

[10] National Research Council, *Proliferation Concerns: Assessing U.S. Efforts to Help Contain Nuclear and Other Dangerous Materials and Technologies in the Former Soviet Union* (Washington, D.C.: National Academy Press, 1997), p. 57.
[11] *MPC&A Strategic Plan*, p. 3.

reports of guards leaving their posts to search for food. Not surprisingly, the guards had little incentive to carry out their duties, and their superiors were not prepared to force the guards to suffer unreasonable hardships. This is a serious concern because the physical protection systems are not effective if guard forces are unavailable to respond to intrusions. The emergency measures of DOE to address these problems during the winter of 1998–1999, undertaken at a cost of about $600,000, were a necessary start in ensuring that the guards could perform at a professional level despite economic hardships.

Many government officials, managers, and workers who have access to direct-use material (or who could arrange such access) have been confronted with economic shortfalls even more severe than those in the dreary days of the early 1990s. As they struggle to keep food on the table, the likelihood of attempted thefts or diversions of direct-use material has increased significantly, according to both U.S. and Russian experts.[12]

ELEMENTS OF A MODERN MPC&A SYSTEM

MPC&A systems are intended to protect material against theft or diversion and to detect such events if they occur.

Physical protection systems should allow for the detection of any unauthorized penetration of barriers and portals, thereby triggering an immediate response, including the use of force if necessary. The system should delay intruders long enough to allow for an effective response. Fences, multiple barriers to entry, limited access points, alarms, and motion detectors are examples of elements of a modern system.

Material control systems should prevent unauthorized movement of materials and allow for the prompt detection of the theft or diversion of material. Such systems may include portal monitors and other devices to control egress from storage sites; authorized flow paths, storage locations, and secure containers for material; and seals and identification codes that make it possible to verify readily the location and condition of material.

Material accountability systems should ensure that all material is accounted for, enable the measurement of losses, and provide information for follow-up investigations of irregularities. Inventory systems and equipment to measure the types and quantities of materials in given areas are important.

Personnel reliability, ensured through security screening, indoctrination, and training, is common to all of the systems. Procedural controls, such as the

---

[12] See, for example, Department of Energy, "Emergency MPC&A Sustainability Measures," November, 1998; Bill Richardson, "Russia's Recession: The Nuclear Fallout," *The Washington Post,* December 23, 1998; and Kenneth Luongo and Matthew Bunn, "A Nuclear Crisis in Russia," *The Boston Globe,* December 29, 1998.

TABLE 1-2 Components of an MPC&A System

	Physical Protection	Control	Accounting
Detection and assessment (sensors, alarms, and assessment systems such as video)	X	X	
Delay (barriers, locks, traps, booths, active measures)	X	X	
Response (communications, interruption, neutralization)	X		
Response team	X		
Entry-and-exit control (badges, biometrics, nuclear material detectors, metal detectors, explosive detectors)	X	X	
Communications and display	X	X	
Measurements and measurement control (weight volume, chemical analysis, isotopic analysis, neutron, gamma, calorimetry)		X	X
Item control (barcodes, seals, material surveillance)		X	
Records and reports			X
Inventory		X	X
Integrated planning, implementation, and effectiveness evaluation	X	X	X
Supporting functions (personnel, procedures, training, organization, administration)	X	X	X

Source: NRC Report, *Proliferation Concerns*

two-man rule (no single employee is left alone in a sensitive area), also play a role.[13]

MPC&A systems rest on the principles of graded safeguards (level of protection is commensurate with the risk of loss of material) and defense in depth (redundant layers of protection). The systems should be sufficiently robust to accommodate threats of all types; threats may be external, such as break-ins by dissident or terrorist groups, or internal, such as thefts by employees. Table 1-2 outlines the basic features of a modern MPC&A system in more detail.

DOE'S COOPERATIVE PROGRAM IN MPC&A

Cooperative efforts to upgrade MPC&A systems in Russia were first considered by the two governments in 1992, but joint projects began only in 1994 because of delays in intergovernmental negotiations. In January 1995, an existing agreement between DOD and MINATOM was amended to add $20 million from Nunn–Lugar funds for MPC&A upgrades, with DOE having the responsibility on the U.S. side.[14] In the meantime, in April 1994, DOE had initiated a second approach that encouraged DOE laboratories to cooperate directly with Russian institutes—the lab-to-lab program.

In a 1995 joint statement, Presidents Clinton and Yeltsin reaffirmed the commitments of the two governments to cooperation in MPC&A, and expanded cooperation followed. Since that time, there have been many U.S.–Russian meetings at the presidential, vice-presidential, and ministerial levels to confirm previous understandings, reach new agreements for specific activities, and reduce impediments to cooperation. These high-level meetings have been followed by dozens of working-level meetings in Moscow and Washington to develop details of the program. DOE has working arrangements not only with MINATOM, but also with the Russian navy, GOSATOMNADZOR, the Murmansk Shipping Company, and a large number of institutes within and outside the MINATOM system.

U.S. funding commitments to the program are set forth in Table 1-3. Although it was envisioned that the program would ramp downward beginning in FY 1999 and into future years, the new recognition of the scope of the problem and the economic downturn in Russia have resulted in a change of U.S. policy. DOE now is committed to a longer-term program, as reflected in statements in January 1999 of both President Clinton and Secretary Richardson expressing strong support for the program.[15]

---

[13] National Research Council, *Material Control and Accounting in the Department of Energy's Nuclear Fuel Complex* (Washington, D.C.: National Academy Press, 1989), pp. 38-42.

[14] The Nunn-Lugar Program dealt broadly with the reduction of the nuclear threat, encompassing weapons dismantlement and storage.

[15] President William J. Clinton, "State of the Union Address," Washington, D.C., January 19, 1999; and Secretary of Energy Bill Richardson, "Remarks at the 7th Carnegie

TABLE 1-3 Finances of the MPC&A Program, Budgeted (Actual) Costs (millions of dollars)

Agency	1993	1994	1995	1996	1997	1998	1999**	2000
DOE	3.0	4.0	12.0	85.0	112.6	132.0	140.1	145.0
	(1.7)	(4.0)	(10.4)	(31.0)	(84.3)	(132.6)	(37.1)	
DOD	0	6.5	60.5	14.3	3.3	0.0	0.0	
	0	(0.7)	(13.5)	(27.9)	(19.2)	(16.7)	(4.4)	
TOTAL	3.0	10.5	72.5	99.3	115.9	132.0	140.1	
	(1.7)	(4.7)	(23.9)	(58.9)	(103.5)	(149.3)	(41.5)	

* The difference between the amount allocated and actual costs is the result of DOE accounting rules on when funds are considered spent. There is a delay of many months between decisions to spend funds on specific activities and the recording of funds as actually spent. So, the amounts for actual costs for each year include funds from previous years.
** The 1999 costs are through January 1999 only.

Source: Department of Energy

Since 1997, the program has been managed on the U.S. side by an MPC&A Task Force within DOE headquarters, which works in coordination with the national laboratories. The government-to-government and the lab-to-lab programs were merged because they had the same objectives, used similar technical approaches, and involved many of the same U.S. specialists.

From the outset, the stated objective of the DOE program has been "to enhance, through Russian–U.S. technical cooperation, the effectiveness of MPC&A in Russian nuclear facilities that process or store HEU or plutonium."[16] The long-term goal is for Russia to support the continued operation of upgraded MPC&A systems at the national and site levels in order to ensure the security of all weapons-usable material within its borders.[17]

DOE initially utilized both horizontal and vertical approaches to address the problems at specific sites. The horizontal approach responded to a common need at many facilities (e.g., portal monitors), and the vertical approach

International Non-Proliferation Conference," Carnegie Endowment for International Peace, Washington, D.C., January 12, 1999.

[16] Joint U.S.–Russian MPC&A Steering Group, "Unified U.S.–Russian Plan for Cooperation on Nuclear Materials Protection, Control, and Accounting (MPC&A) Between Department of Energy Laboratories and the Institutes and Enterprises of the Ministry of Atomic Energy (Minatom) Nuclear Defense Complex," September 1, 1995, p. 5.

[17] Joint U.S.-Russian MPC&A Steering Group, "Unified U.S.-Russian Plan for Cooperation on Nuclear Materials Protection, Control, and Accounting Between Department of Energy Laboratories and the Institutes and Enterprises of the Ministry of Atomic Energy Nuclear Defense Complex," September 1, 1995, p. 5.

concentrated on installing complete systems at selected facilities. More recently, DOE has emphasized the vertical approach. The general Program Guidelines issued by the Task Force in January 1998 are set forth in Appendix E. These guidelines broadly sketch the program mission, from establishing MPC&A cooperation at all sites to implementing systematic and rapid upgrades to ensuring sustainability. The Task Force also issued a guidance document concerning the approaches that are to be implemented at specific sites.[18]

DOE believes that it has initiated program activity at almost all sites where direct-use material is located. However, at very few sites has the program involved activities at all buildings where material is stored, and at many sites, the contents of some buildings are known only in very general terms. DOE's imperfect knowledge and some limitations on the scope of the MPC&A programs are inevitable because the program is concerned with activities at the core of Russian national security activities. Overall, the program has been reasonably successful in overcoming Russian concerns as to U.S. motives, although lingering suspicions probably remain among some Russian officials.

Finally, the MPC&A program is only one of several national security programs in Russia supported by DOE. Other programs are the Nuclear Cities Initiative (to encourage commercial activities in closed cities), the Initiatives for Proliferation Prevention (to provide appropriate civilian-oriented employment opportunities for former weapons scientists), cooperation on nuclear reactor safety, the U.S. purchase of 500 tons of HEU from Russia, experiments with MOX reactor fuel, conversion of the nuclear reactor cores in power plants at Tomsk 7 and Krasnoyarsk 26, coordination of activities related to nuclear smuggling, and the broader programs on disposition of excess plutonium.[19] In addition, there are other related programs of the U.S. government, including the projects of the International Science and Technology Center, the Cooperative Threat Reduction program of DOD, and the assistance efforts of the U.S. Agency for International Development. As DOE's MPC&A Task Force readily acknowledges, there has not been sufficient cooperation among these activities to ensure that they reinforce one another.

---

[18] *Guidelines for Material Protection, Control, and Accounting Upgrades at Russian Facilities*, December 1998.

[19] For an overview of many of these programs, see Matthew Bunn and John Holdren, "Managing Military Uranium and Plutonium in the United States and the Former Soviet Union," *Annual Review of Energy and the Environment*, Vol. 22, pp. 403-486, 1997.

2

FINDINGS AND RECOMMENDATIONS

GENERAL FINDINGS

1. A National Security Imperative

As discussed in Chapter 1, the recent decline in the Russian economy has severely affected many Russian government officials, nuclear specialists, and workers who have access or could arrange access to direct-use material. The economic deprivation has increased the likelihood of attempted thefts or diversions of such material from Russian facilities. Meanwhile, expanded access by U.S. specialists to Russian facilities has led to increased estimates of the number of buildings where direct-use material is located and of the effort that will be required to install adequate material protection, control, and accountability (MPC&A) upgrades throughout the Russian nuclear complex. Experience also has led to longer and more realistic timelines for overcoming administrative and technical problems in installing upgrades and has underscored the problems that will be encountered in maintaining them after they are in place.

At the same time, there are many examples of impressive progress directly attributable to U.S. efforts (e.g., consolidation of material into a limited number of buildings at Luch, construction of security-enhanced railcars, initiation of the naval fresh fuel program). Indeed, in the absence of the U.S.-financed program, the situation undoubtedly would be far more dangerous. The First Deputy Minister for Atomic Energy stated in January 1999 that U.S.–Russian cooperation has been an important factor in strengthening Russian security efforts at many facilities.[20] U.S. involvement has been pivotal in stimulating Russian efforts, though limited, to develop a stronger indigenous capability for installing and maintaining MPC&A systems and in raising awareness of the importance of MPC&A throughout the Russian nuclear complex.

---

[20] Lev Ryabev, "Remarks at the 7th Carnegie International Non-Proliferation Conference," Carnegie Endowment for International Peace, Washington, D.C., January 11, 1999.

The Department of Energy's (DOE's) original plan was to reduce its involvement in Russia after 1998 and to have initial MPC&A upgrades completed at all facilities with direct-use material by 2002. In light of the decline in the Russian economy, the more recent estimates of the extent of the MPC&A problems, and the delays encountered in installing MPC&A systems, this schedule is now completely unrealistic. The need for a longer-term U.S. effort is clear.

There is a strong U.S. national security imperative for substantial U.S. involvement in MPC&A projects in Russia for at least the next decade. Meanwhile, the U.S. government must continue to emphasize the importance of MPC&A as a nonproliferation imperative at the highest political levels in Russia in order to achieve the final goal of ensuring that MPC&A systems are in place and operating effectively at all locations and are supported financially by the Russian government.

2. Program Progress in Key Areas

There are several key elements of the MPC&A program that deserve continuing attention.

Physical Protection: Considerable progress has been achieved in installing physical protection systems, but the level of protection at various sites is quite uneven. At some buildings, state-of-the-art systems are fully operational, whereas at others, the systems have not been well designed or are not working as intended. Once upgrades have been installed at a building or set of buildings, DOE typically organizes a well-publicized commissioning ceremony that may give the false impression that the job is done. These ceremonies are not always understood by Russian and U.S. officials as simply signifying that important progress has been made; and too little attention has been given to ensuring that upgrades are operated and maintained as intended.

Accountancy: Little progress has been made in upgrading the primitive material accountancy systems used at almost all sites and at the national level. Accountancy systems must be an integral part of an overall system to protect direct-use material, particularly from insider threats. If up-to-date information on precisely what material is on site and where it is located is unavailable, then it is not possible to determine whether material is missing. Although many years will be required to complete this task, more aggressive efforts clearly are warranted. Unfortunately, progress in installing upgraded accountancy systems has not been used as a metric by DOE to measure the success of its efforts.

There is considerable accountancy-related activity at some sites, including the installation of many computers and computer programs as well as measurement instrumentation. But low priority has been given to carrying out and completing adequate physical inventories of materials and maintaining inventory balances on a continuing basis. The importance of the initial inventory as a baseline for the accountancy system has been highlighted at Kurchatov Institute where, despite a comprehensive paper accountancy system, numerous errors in the records have been discovered during recent inventories in several buildings. At the national level, organizational and technical uncertainties continue to plague the development of a national accountancy system. In short, neither U.S. nor Russian senior officials have given accountancy adequate attention.

Consolidation: At a few sites there has been progress in consolidating direct-use material into fewer buildings. However, there has been no progress in reducing the number of sites within Russia where material is located, and the outlook for such intersite consolidation is not bright given the interest of each facility in retaining its right to have direct-use material. Nonetheless, consolidation among and within sites is essential to reduce the technical problems and to minimize the costs in ensuring the security of all direct-use material on a national scale. Given the incentives at the facility level in maintaining an inventory of direct-use material (if only to participate in and receive funds from U.S. programs), consolidation across sites will not occur without high-level pressure and economic incentives from the U.S. and Russian governments.

Access: DOE has been quite successful in gaining access for U.S. specialists to most sites where direct-use material is believed to be located. Development of mutual trust at both the official and working levels has been an essential aspect in expanding activities into sensitive facilities. However, years will be required at some sites to build such trust and to gain even limited access to all buildings where significant quantities of material are located. Indeed, access to extremely sensitive buildings by U.S. officials probably never will be achieved. The program should take into account this reality. Of special relevance, DOE has successfully relied on well-qualified colleagues from Kurchatov Institute in order to initiate the naval fuel program, which involves access to sensitive facilities. Reliance on qualified intermediaries may succeed at other sensitive facilities as well, such as the serial production facilities (facilities where warheads are assembled and disassembled).

Neglected Material: DOE has concentrated almost exclusively on protecting unirradiated HEU and separated plutonium. Although the focus on such direct-use material is appropriate, there are also large quantities of spent fuel from maritime, research, and breeder reactors that are inadequately protected. Fuel with low burnup rates and/or long storage times is not "self-protecting" and also may pose serious proliferation threats. DOE, along with other U.S. agencies, is participating in cooperative programs to provide interim storage for spent naval fuel, particularly on the Kola peninsula. However, it appears that inadequate attention is being given to the MPC&A aspects of fuel elsewhere that, if stolen or diverted, could be processed for weapons use.

Testing the System: The development of a high-quality MPC&A system involves testing the system, fixing the weaknesses revealed by the test, and testing again. There is no national program for realistic testing of MPC&A systems in Russia. Those tests that have been conducted by U.S. teams have identified flaws in "completed" MPC&A systems, suggesting that, just as in the United States, repeated tests are necessary to have the systems work well. DOE has not insisted that a comprehensive testing program be put into place.

3. **Russian Interest in the Program**

Russian support for the program continues to be strong, as evidenced by public statements by senior Russian officials; by agreements with the Ministry of Atomic Energy (MINATOM), the navy, and other organizations; and by U.S. access to a large number of facilities that only a few years ago were unknown to the outside world. At the same time, Russian motivations for participation in the program are mixed.

Opportunities for financial support are welcomed at both the governmental and institute levels in Russia. At one extreme, there have been reports that some Russian institutes will agree to any approach advocated by U.S. specialists if it is accompanied by U.S. funding. Other reports suggest that at some locations Russian MPC&A teams work hard in anticipation of visits by the U.S. teams, but then slack off until the next opportunity for contracts arise. Overall, however, the Russian performance in carrying out contractual obligations seems to be very good, and at many sites the return on the U.S. investment is high.

At the level of the Russian government, there are obvious foreign policy benefits from participation in the program at a time when the porosity of the Russian nuclear complex is a continuing concern. Moreover, there is a growing cadre of Russian specialists at the institute level who clearly are committed to establishing and operating high-quality MPC&A programs. Many have a full appreciation of the importance of nonproliferation goals. They, along with others who are less concerned with international security, also are driven by professional pride.

It appears, however, that many Russian institute leaders are less concerned with the inadequacies of existing MPC&A systems, and particularly with the need for vigorous efforts to counter *insider* threats, than are U.S. specialists. Russian managers often seem more concerned about the penetration of facilities by *outsiders* intent on sabotage or theft of items that can be sold easily on local markets, rather than about internal theft or diversion of direct-use material. Thus, whereas U.S. specialists emphasize protection of direct-use material as close to the source as possible, the Russian starting point for protecting the assets of an institute, including its direct-use material, is usually the installation of perimeter fencing adequate to enable the guard force to keep unauthorized personnel off the premises. The differing perception of the threat—and of the optimal means to address it—results in challenging problems in designing, installing, and maintaining systems that meet both U.S. and Russian objectives.

Formal MPC&A training programs that have been established in Russia through DOE cooperative efforts seem to be well designed and organized and quite popular among Russian officials, administrators, instructors, and students—at least as long as the programs are subsidized. These programs have been quite effective in raising the level of MPC&A competence and awareness in Russia. However, the cost per student attending the graduate-level program at Moscow Engineering Physics Institute (MEPhI) is high, and the short courses at Obninsk are dependent on a large number of foreign instructors. Also, the Obninsk programs are oriented heavily toward the technological aspects of MPC&A with minimal attention to the need to ensure that students are familiarized with nonproliferation issues as well as with safety and theft-prevention dimensions of MPC&A systems.

There are also opportunities for on-the-job training at Russian facilities. Ministry of Interior (MVD) guards and junior professional employees participate in this training, but they also could benefit from formal programs that enhance understanding of nonproliferation goals of the program as well as the technical aspects of MPC&A systems.

4. **DOE's Management of the Program**

The recent interest of DOE Secretary Richardson in the program,[21] after an apparent decline of active high-level involvement within DOE during 1997 and 1998, is a welcome development. The Department of State and the Department of Defense (DOD), as well as DOE, support programs to reduce the likelihood of leakage of nuclear materials and technology from Russia to countries of concern. Any weakening of DOE's commitment to the MPC&A program, which is a cornerstone of all of these activities, will undermine the overall effort to reduce the dangers of nuclear proliferation.

Also, as noted in Chapter 1, coordination among the programs managed by DOE is important, for example, MPC&A, highly enriched uranium (HEU) purchase, Initiatives for Proliferation Prevention (IPP), Nuclear Cities Initiative, and plutonium disposition. At times, different DOE laboratories with redundant interests and capabilities are involved in related, but uncoordinated, programs at the same Russian sites. And, on occasion, DOE laboratory participants are not aware of overlapping DOE activities that are under way or that could be initiated to complement their MPC&A efforts.

DOE has taken steps to correct earlier coordination problems within the MPC&A program itself. There has been considerable progress in improving the internal flow and consistency of program documentation and in sharing information among laboratories and headquarters units. However, in the process of gaining better control over a rapidly expanding program, DOE has established additional levels of line management within DOE headquarters. There are also examples of micromanagement of technical activities by

---

[21] See, for example, Secretary of Energy Bill Richardson, "Remarks at the 7th Carnegie International Non-Proliferation Conference," Carnegie Endowment for International Peace, Washington, D.C., January 12, 1999.

headquarters personnel that should remain the province of skilled personnel at the laboratories.

The actual implementation of MPC&A programs at individual Russian facilities should be the responsibility of site managers. Site managers, who are drawn from DOE's laboratories, should have primary responsibility for developing and overseeing implementation of the site's workplan, which includes the specific MPC&A requirements and the schedule for completing upgrades. Overall, the MPC&A site managers appear to be well qualified and to have done commendable jobs of installing new physical protection systems. However, there are examples of inadequate oversight of implementation by laboratory personnel in the field. Also, efficiencies could be achieved by more careful selection of team members who travel to Russia both to reduce redundant skills and to eliminate "observers" who sometimes travel simply to maintain a presence of a DOE laboratory on a team.

Finally, although there has been significant progress at a number of Russian sites and serious problems at others, there is no institutionalized means for either U.S. or Russian participants in MPC&A programs to share lessons learned with colleagues. The annual meeting of the Institute for Nuclear Materials Management provides a useful forum to consider broad issues, but it does not provide opportunities for the more detailed discussions that are needed. The absence of lateral communication among MPC&A directors as well as among senior personnel from Russian institutions is especially acute.

SPECIFIC FINDINGS AND RECOMMENDATIONS

1. Sustain the U.S. Commitment to the Program

Finding:

Continued U.S. support of the program is necessary to ensure that needed upgrades are installed promptly in hundreds of buildings at many sites, that the systems that are installed are operated and maintained as intended, and that guard forces stay on the job. The committee is aware of only two programs for which sustained Russian financial support of MPC&A personnel and activities seem highly likely—namely, the navy fresh-fuel program and the MPC&A program at the Joint Institute for Nuclear Research in Dubna, which will be funded, at least in part, by the German government's annual contribution to the overall activities of the institute.

The Department of Energy has requested $145 million for FY 2000, a level about the same as expenditures scheduled for FY 1999. DOE program managers have informally advocated this level of funding for the next five years, although DOE has not yet adopted a position on funding beyond FY 2000. The committee believes that the informal five-year projection is realistic, given the limited capacity of Russian institutions to use funds effectively. At the same time, however, there may be new opportunities to use additional funds effectively. In any event, DOE needs to promptly complete its current effort to conduct a comprehensive baseline study of the needs for all known buildings of relevance in Russia, with appropriate estimates of the costs that would be entailed to complete the overall effort. This study should provide a firmer base for budget projections.

Recommendations:

A. Maintain the current level of U.S. support ($145 million per year) for the program for at least the next five years and be prepared to increase funding should particularly important opportunities arise. Plan to continue an appropriately scaled program of cooperation thereafter, with the scope and duration of the program depending on progress in installing MPC&A upgrades and economic conditions in Russia.

Given the seriousness of the threat, current levels of support must be maintained. Furthermore, new opportunities, particularly activities involving spent fuel and intersite consolidation, will require additional funding as they arise. There is little likelihood over the next several years that the Russian government or the institutes will have funds to continue many aspects of the MPC&A program on their own. U.S. national security interests provide a compelling reason to continue the current level of U.S. funding during the economic turmoil in Russia.

The U.S. government should, of course, continually press the Russian government and the individual institutes to finance as much of the program as possible. However, even when economic conditions improve and funding becomes less constrained, continued U.S. involvement in cooperative MPC&A endeavors should encourage the Russian government to adhere to its commitment to upgrade MPC&A systems and the institutes to devote their own funds to the support of MPC&A specialists and equipment.

B. Provide support for operational costs of selected aspects of the personnel and technical infrastructure at Russian institutes to help ensure that MPC&A systems that have been installed are operated and maintained as intended.

Implementing and maintaining MPC&A systems as they were designed is just as important as the installation of sound systems in the first place. If there are communication or equipment failures, if the electricity at a site is disrupted because of payment arrears, if specialists are distracted by the need to obtain supplemental income from other activities simply to survive, and if guards are not at their posts because there are no coats for outside patrols or they must search for food, the protection provided by investments in technical systems will be reduced substantially.

The committee recognizes that the recent emergency measures of DOE have addressed some of these problems, particularly the needs of guards. However, unless economic problems subside, DOE should be prepared to provide greater support for operational activities.

C. Ensure that projects for the development and operation of MPC&A systems, as well as associated training programs, include opportunities for participation by Russian guards.

Guards remain a critical aspect of physical protection systems at almost all sites in Russia, both because of the security they provide and the possibility that, if faced with extreme economic hardships, they could themselves become a source of insider threats. At present, deterioration of morale and of inattention to duty appears widespread within Russian security forces. Modest investments to help ensure that guards involved in MPC&A are rewarded for doing a good job should become an important component of the effort to contain direct-use material.

This support should be structured so that it does not simply subsidize Russian security agencies. Financial benefits should be coupled with requirements for the guards to upgrade their skills through attendance at training programs and with tests of their capabilities to respond to simulated penetrations of facilities. The temporary nature of the associated financial benefits should be very clear from the outset because Russian security forces probably will be among the early beneficiaries of an economic recovery. Nevertheless, to ensure the long-term professionalism of the guards, carefully designed training programs for guards will need to be continued.

D. Encourage both the Russian government and institutes to seek additional income sources for supporting MPC&A programs.

Developing funding sources for any activity in Russia is a formidable challenge, but Russian resources are critical to sustaining the program in the long term. In the immediate future, even limited funds devoted to MPC&A from a variety of sources could send an important signal to all participants as to the priority of this activity. Russian income from the sale of HEU to the United States should be considered as one source of funds dedicated to MPC&A. Also, there may be opportunities for the International Atomic Energy Agency, EURATOM, and other international programs and for bilateral programs of European governments to become involved to a greater extent. To this end, the current MPC&A upgrading activities should shed their image as being almost entirely a U.S.–Russian bilateral program. However, in the near term, it is unlikely that other countries will shoulder much of the financial burden, and the U.S. commitment should not be reduced in anticipation of significant foreign contributions that may not materialize.

2. Reassess Priorities to Address Important Vulnerabilities

Finding:

Activities currently under way are largely the result of addressing "targets of opportunity." Although many activities are directed to high-priority concerns, inadequate attention has been given to systematic targeting of the most important nationwide vulnerabilities. Furthermore, on occasion, the costs of the advanced technological systems that would best address these vulnerabilities are simply too great under current budgetary constraints, and action is deferred without considering less expensive interim steps. Related to these concerns, the Minister of Atomic Energy informed the committee of his interest in having an "integrated system" and of his apprehension that many individual activities are not tied together in a rational way. With these considerations, the committee identified some key priorities that need greater attention.

Recommendations:

A. Review the languishing materials accountancy programs at all sites and, as part of adjusting overall program priorities, devote additional resources to improve and speed up performance in this area.

U.S. officials and specialists must impress on Russian colleagues the importance of knowing at all time the whereabouts of all direct-use material—classified by type and quantity—and the personnel responsible for the material. The accountancy system must be able to detect discrepancies between expected and actual material inventories. In the absence of such an accountancy system, the diversion of material may remain undetected. Financial incentives to encourage a more serious Russian "buy-in" to this concept are needed. These incentives might include bonus clauses for superior Russian performance in statements of work. Implementation of these systems should begin as soon as possible at each site; there is no need to wait for a national accountancy system to be developed. Although developing a complete inventory at each site may take considerable time, as a first step all items or containers with direct-use material should be located, counted, logged, and sealed. The actual measurements of the quantity could be a second step. Schedules and milestones for each site must be developed, monitored, and given the highest priority.

B. Give greater priority to developing an appropriate national material accountancy system, ensuring that different types of accountancy systems being installed at individual facilities have the capability to provide data in a form that can be incorporated easily into the national system.

A well-developed and vigorously enforced national system will help to ensure that site-level systems are established, maintained, and operated and that materials being transferred between sites are adequately monitored and controlled. Many years will be required to achieve the goal of a high-quality national system, and basic decisions on the character of the system are needed promptly. Adjustments in overall priorities should ensure that the relevant organizations are provided with adequate resources to meet their responsibilities.

C. **Continue to consolidate storage areas for direct-use materials whenever possible and give greater attention to the establishment of well-designed central storage facilities that serve more than one site**.

DOE's emphasis on the importance of consolidation, highlighted in the March 28, 1999, DOE–MINATOM agreement on this topic, should continue. It is clear that intrasite consolidation is an important first step, and information about successful consolidation efforts at some sites, including projected long-term savings in MPC&A costs, should be disseminated widely. Despite likely near-term opposition from institutions that are determined to maintain stocks of direct-use material regardless of current needs for the material, DOE should continue to push for intersite consolidation and be prepared to take swift advantage of opportunities as they arise. As a step in this direction, DOE should encourage institutes at sites with inadequate storage facilities to use high-quality central storage facilities at other sites as repositories for material in long-term storage.

D. **Recognizing Russian security constraints, develop as complete an MPC&A plan as possible for each site where there is direct-use material.**

Preparation of sitewide plans are an important step in addressing the most vulnerable material at the sites. Now that DOE has established its presence at most sites and there are well-developed guidelines for DOE MPC&A upgrades at Russian facilities, there should be increased attention to expanding the coverage of site plans that are currently inadequate. Given the sensitivity of this topic, these site plans should be developed with intensive involvement of Russian counterparts. Indeed, it often may be appropriate and necessary to ask Russian counterparts to take the lead in this activity in light of Russian national security sensitivities.

E. **Establish MPC&A programs at the serial production facilities as soon as possible.**

DOE should continue its efforts to engage the serial production facilities, where warheads are assembled and disassembled, in the program. These sites are very sensitive and the Russian hesitation about opening them to the United States is understandable. DOD encountered similar problems in addressing security improvements for nuclear weapons, and DOE–DOD exchanges of experiences on successful and unsuccessful approaches should be encouraged.

F. Develop programs to address icebreaker and naval spent fuel that are of proliferation concern.

Initiatives in the marine area should build on the successes to date in developing MPC&A systems for naval fresh fuel. Early steps to identify the amount, general characteristics, and location of maritime spent fuel have begun. Opportunities to integrate DOE's MPC&A objectives with the Russian spent fuel program should be pursued. Marine-oriented programs might be most effectively handled as a discrete cluster of activities because many of the same officials and specialists from both countries likely will be involved.

G. Develop programs to address spent fuel from reactors other than naval reactors.

The extent to which spent fuel from other types of reactors poses a proliferation threat needs detailed investigation. Of special concern are the plutonium reactors at Tomsk and Krasnoyarsk, the breeder reactor at Beloyarsk, and a number of research reactors. Because much of the spent fuel is quite old, there may be storage areas that contain materials of proliferation concern.

H. Expand the transportation program to provide a larger number of more secure trucks to a variety of facilities while ensuring the soundness of the procedures for tracking the movement of direct-use material.

Although a good start has been made in constructing railcars and trucks for transporting material between sites, many more trucks clearly are needed for intersite and intrasite transport. Considerable direct-use material is being transported because of new programs on warhead dismantlement, uranium sales, core conversion programs, and plutonium disposition as well as continuing programs for refueling maritime nuclear reactors. Precisely how many vehicles are needed should be determined by a careful analysis of transport requirements. Also, in light of concerns about material accountancy systems, the procedures for monitoring and controlling intersite shipments of material need careful review.

I. Recognize that in the near term, because of economic and other factors, it may be necessary to install systems that fall short of internationally accepted standards, in anticipation of subsequent refinements. In this regard, use appropriate MPC&A measures whether they involve high-technology or low-technology approaches.

Fences, padlocks, and other low-technology approaches may not be an adequate solution for long-term containment of direct-use material. However, in some cases, such systems can provide a degree of interim protection. Similarly, the development of accurate and reliable accountancy systems should not be postponed until advanced computer systems are installed. Because limited funds always will be a constraint, low-cost temporary measures should be considered if a situation needs immediate attention.

3. Indigenize MPC&A Capabilities

Finding:

U.S. specialists have played the lead roles at most sites where cooperative MPC&A activities have been undertaken. Although there are many examples of steps being taken to shift more responsibilities to Russian counterparts for the design and installation of systems, the process of indigenization of activities should receive still higher priority. This process is crucial to the proper functioning of MPC&A systems both in the immediate future and in the long term. Also, given the difference in salary requirements of U.S. and Russian specialists, such a shift will permit the stretching of available financial resources across a broader spectrum of activities.

Recommendations:

A. Increase the percentage of available U.S. funding that is directed to financing activities of Russian organizations, with a steadily declining percentage directed to supporting U.S. participants in the program.
The current division of funding between support of Russian and support of U.S. institutions is about 50-50.[22] As Russian specialists increase their capabilities to take on more of the responsibility for the program, there are opportunities for cutting back on the demand for involvement of U.S. specialists, some of whom are approaching burnout, simply by allowing Russian entities to play larger roles. These entities might include well-qualified Russian firms or U.S.–Russian joint ventures capable of implementing MPC&A. This new emphasis also should reduce costs because of salary differentials between U.S. and Russian specialists.

B. Expand efforts to utilize Russian equipment and services whenever possible and to encourage Russian enterprises and institutes to increase capabilities to provide high-quality equipment and associated warranties and services.
In the long run, a strong indigenous industrial capability will be essential for sustaining systems in Russia. This capability will develop only if there is a demand for locally produced products. DOE specialists should continue to work with a variety of Russian enterprises and institutes and, in cooperation with Russian counterparts, bring them into the program as they demonstrate satisfactory capabilities. In cases in which Russian manufacturers are having difficulty achieving acceptable international performance standards, they might be encouraged to enter into licensing arrangements with foreign suppliers. The

---

[22] According to DOE, approximately one-half of the money supporting Russian institutions is for equipment, most of which is purchased in either Russia or the United States. DOE does not have data readily available on how much of this equipment is purchased in Russia.

committee recognizes the problems that have been encountered in DOE's efforts to use Russian-produced equipment and DOE's limited ability to address these problems without substantial cooperation by the Russian counterparts. However, DOE should not lose sight of the need to encourage high-quality Russian equipment and service as a long-term goal.

C. Use Russian specialists from institutions with well-developed MPC&A capabilities to replace some U.S. members of teams at Russian institutions with less developed capabilities.

In connection with the program for containing naval fresh fuel, specialists from the Kurchatov Institute are setting an important example by demonstrating that qualified Russian experts can become trainers of Russian colleagues, as well as advisers on the detailed aspects of MPC&A systems. Specialists from other institutes that have achieved significant transformations in their approaches to MPC&A, such as Luch, also could serve as important members of teams assembled by DOE to assist with upgrading MPC&A systems at Russian facilities that have entered into the program only recently.

D. Rely increasingly on Russian specialists to replace U.S. specialists in presenting MPC&A training programs at Obninsk and other training sites.

Most aspects of the MPC&A courses offered at the training centers in Obninsk could be presented by Russian specialists, including some who are affiliated with nongovernmental organizations. Reliance on local expertise has several advantages: instruction in Russian, greatly reduced costs for instructors, and opportunities for the instructors to improve their own expertise through teaching.

E. Encourage MEPhI to increase student participation (and its income resulting from tuition payments) in its security-oriented courses by offering an industrial security as well as an MPC&A specialization.

The 18-month program in MPC&A studies offered by MEPhI, although very impressive for specialists in the field, may be too narrow in scope to be sustained financially in the long term. The MPC&A course has attracted considerable attention from the industrial security community in Russia and could become a cornerstone of a broader security curriculum that would provide a high likelihood of employment opportunities for graduates. MPC&A studies could continue as one specialization within a broader set of course offerings that also would attract security specialists in demand by Russian banks and industrial enterprises. The program also might be of interest to students from other countries in the region.

F. Give greater attention, in both training and implementation activities, to developing personal commitments on the part of Russian managers, specialists, and guard forces to fulfill their responsibilities for ensuring the proper functioning of MPC&A systems.

Although modern MPC&A systems are highly dependent on technological devices, they are only as effective as the people operating them. Motivating managers, specialists, and guard forces to be enthusiastic or even attentive to duty during a time of economic crisis is difficult. Periodically at each site, DOE should recognize good performance by Russian participants with plaques, certificates, and financial rewards. In addition to financial incentives to encourage a greater sense of responsibility among the participants, training and implementation activities should emphasize the importance of MPC&A activities and the national security consequences if the systems are breached. Furthermore, MINATOM training institutes should be encouraged to incorporate more MPC&A issues and courses into their curricula. Also, all DOE participants in cooperative activities should highlight the centrality of an MPC&A ethic that requires the reporting of violations and that does not tolerate shortcuts or exceptions in implementing MPC&A systems.

G. Increase opportunities for Russian input in establishing priorities at specific sites and in preparing statements of work for individual projects.

The greater the degree of Russian "ownership" of the upgrades that are being put in place, the more likely is the Russian enthusiasm for expediting their introduction and for ensuring that they operate as intended. Because the U.S. side controls the funding and usually takes the lead in preparing all contract documents, it is often difficult for Russian specialists to feel that they are equal partners in project design. The initial drafts of joint statements and workplans might be prepared in Russian to ensure that the Russian views are adequately represented and jointly reviewed and modified as appropriate. In general, U.S. site managers need to take whatever time is necessary so that Russian views receive weight comparable to U.S. views and that the designs that emerge are truly joint designs, both in perception and in fact.

4. <u>Reduce Impediments to Effective Cooperation</u>

<u>Finding</u>:

Progress in upgrading MPC&A systems has been delayed by administrative problems encountered at the national and facility levels, such as uncertainties as to participation by Russian institutions, access to sensitive facilities, lack of understanding as to tax and customs issues, confusion as to certification requirements for equipment that is to be used, and Russian indecision concerning the national materials accountancy system. Also, progress is related directly to financial incentives for the participating Russian institutes, and DOE has not taken advantage of resources available through other programs to increase such incentives.

Recommendations:

A. Give higher priority within DOE headquarters to intergovernmental discussions of issues that impede rapid progress.

There are a number of key issues that can be resolved only at the intergovernmental level, which usually means that they must be considered at the DOE–MINATOM level. Examples include procedures in Russia for the certification of equipment to be used in MPC&A systems, exemptions of funds and equipment transferred to Russia from taxes and from customs duties, procedures for access to sensitive facilities, responsibilities in Russia for developing a national system for material accountancy, and the lack of willingness of the managers of sensitive facilities (e.g., Electrostal and the serial production facilities) to participate fully in the program. Such issues should be a top priority of DOE headquarters.

B. Develop an improved political/legal framework for U.S.-funded MPC&A activities in Russia that ensures long-term stability for the program and exemptions from taxes, customs charges, and related fees.

As of March 1999, the U.S. and Russian governments had completed negotiations of a draft agreement to exempt U.S.-funded programs in Russia, such as the MPC&A program, from certain taxes and customs duties. Once this agreement is in place, DOE should take additional steps as necessary with MINATOM and other Russian organizations to ensure that these exemptions are fully honored at all participating institutions.

C. Encourage greater interest in MPC&A at the institute level by providing rewards for good performance in developing and implementing MPC&A programs, such as priority opportunities for participation in other U.S. government-sponsored programs.

The linkages between the MPC&A program and other DOE programs being carried out in Russia (e.g., IPP, Nuclear Cities Initiative, lab-to-lab research projects, support through the International Science and Technology Center [ISTC], plutonium disposition) should be strengthened. These programs can provide substantial resources to Russian institutes, and many institutes are involved in several programs. DOE should manage these programs as complementary efforts so that each reinforces the other. An institute's commitment to and progress in MPC&A should be an important factor when considering that institute's participation in other programs. It seems difficult to justify lucrative contracts in the nuclear field from U.S. government sources for an institute that has a poor MPC&A record.

D. Establish in Moscow a DOE MPC&A office that can troubleshoot and help overcome barriers to rapid progress and that can facilitate the coordination of MPC&A activities with other DOE programs.

Onsite investigations of problems by resident staff often could save considerable time and expense associated with bringing in troubleshooters from the United States. Although it seems unlikely that a posting in Moscow would be of interest to highly qualified technical personnel, the office could program the time of technical experts from the United States in a more efficient manner. Also, given the many diverse and decentralized DOE activities in Russia, a field office would be in a good position to obtain the information necessary for ensuring the coordination of DOE efforts at specific sites. The office should have wide-ranging authority within DOE for the collection and distribution of information, but it should not be empowered to control activities that are the responsibility of headquarters' units or the laboratories.

5. Improve Management of U.S. Personnel and Financial Resources

Finding:

The challenge of managing the activities of a multitude of U.S. laboratories, U.S. contractors, and DOE headquarters personnel at about 50 sites in Russia needs a comprehensive review. Steps should be taken to maximize the return on U.S. expenditures, to reduce redundancies of responsibilities while ensuring adequate oversight, and to provide incentives that will attract highly qualified specialists.

Recommendations:

A. Develop procedures for ensuring that funds transferred to Russia are not subject to taxes, contributions to Russian pension or social funds, or excessive overhead charges.

Some organizations, such as the ISTC and the U.S. Civilian Research and Development Foundation, are provided exemptions by Russian authorities from payments of taxes, customs duties, and contributions to pension and social funds when transferring funds to Russian institutions and individuals. Use of these well established mechanisms seems to be a feasible approach for at least some aspects of the MPC&A program. This approach would not affect the technical aspects of the program and, once established, would not slow down contracting procedures. This approach is being pioneered by Oak Ridge National Laboratory, which has become an ISTC partner, an initiative that should be of priority interest to other laboratories as well. Even if the U.S. government is successful in negotiating an improved overarching legal/political framework that addresses such issues as suggested above, the use of additional channels for transferring funds might provide options should difficulties arise in direct dealings with MINATOM institutes.

B. Whenever possible, reduce the size and frequency of U.S. teams traveling to Russia for negotiations, site visits, and other reasons.
Although frequent visits to Russia by U.S. specialists are important, the number of visits appears excessive (about 100 specialists were traveling to Russia each week at the beginning of 1999). Highest priority should be given to negotiations at the intergovernmental level to resolve issues that impede progress and to technical visits that the site managers consider critical to putting MPC&A systems in place. However, even for these activities, DOE headquarters should ensure that every member of the teams has a clear job to do. As for other types of visits (e.g., participation in training programs, commissioning ceremonies, field audits, conferences), there should be opportunities to reduce travel.

C. Develop a clearer division of responsibility between DOE headquarters staff and specialists of the DOE laboratories. This division should recognize the lead role of headquarters in intergovernmental negotiations, formulation of general policy guidance, determination of priorities among sites, and financial oversight. It should recognize the role of the laboratories in providing advice to headquarters on policy aspects of the program, in making technical decisions in accordance with headquarters guidance and budgetary allocations, and in providing specialists who are responsible for the development and implementation of MPC&A upgrades.
The responsibilities of DOE headquarters for overall design of the program, for coordination, for oversight, and for budgetary justification of the program are clear. However, DOE should recognize the primacy of the laboratories in implementation and refrain from attempting to micromanage activities at individual sites. At the same time, senior DOE personnel should have more direct access to field activities; and to this end, the organizational structure at DOE headquarters needs to be flatter so as to reduce the several layers of management between the Assistant Secretary for Arms Control and Nonproliferation and the site managers.

D. Give greater recognition to the key role of U.S. site managers, ensure that they have the necessary authority to manage all activities at the sites and to make key technical decisions, and design support systems that facilitate rather than impede their activities.
Site managers should be given broad authority to act in accordance with policy guidance and financial resources provided by DOE headquarters. They then should be held fully accountable for results. They should be able to allocate budgeted resources with some flexibility, drawing on DOE expertise across the laboratories as needed. Of course, they must coordinate their activities with many other DOE site managers competing for the same personnel resources. However, once in the field, their authority should be clear. DOE should conduct periodic reviews of the activities of each site manager, and those who do not

produce technically or administratively acceptable results should be replaced without delay.

E. Select specialists for each field activity on the basis of their personal qualifications and availability and abandon previous policies that assigned slots for sites to specific DOE laboratories.

If the site manager is to have responsibility for performance of the site team within budgetary constraints, he or she should have a degree of control over team membership. He or she should not be limited to drawing from specific laboratories or required to pad the team with unnecessary specialists simply to maintain the presence of specific laboratories on the team. Of course, coordination among site managers in the competition for specialists is very important.

F. Ensure that there is a cadre of specialists available for consultations with DOE headquarters, DOE laboratories, and participating contractor organizations who have strong backgrounds in Russian language, history, culture, economics, and accounting and financial systems, as well as in familiarity with day-to-day problems encountered in working in the Russian environment.

All specialists traveling to Russia should have some understanding of the Russian environment. Even those who already have made multiple visits could benefit from additional background concerning the historical and current settings. To this end, DOE headquarters, laboratories, and contractors should arrange for short training programs, lectures, and/or consultations for program participants. DOE's effectiveness would be enhanced if program participants had greater familiarity with Russian traditions, sensitivities, and realities.

G. Develop better metrics for gauging the success of MPC&A upgrades at sites.

Altough considerable efforts have been devoted to the installation of physical protection equipment, the operation and maintenance of this equipment and material accountancy have not received adequate attention. Until there has been substantial progress on material accountancy and on ensuring the sustainability of physical protection equipment, work at sites should not be considered complete. Another metric of progress could be the success of new systems in preventing test penetrations of the facilities.

H. Improve communication in Russia and in the United States between site managers.

Many problems are common to sites across Russia. However, there is no institutionalized mechanism for site managers to share their experiences either in Russia or the United States, to discuss the success or failure of various approaches, and to determine common difficulties that need to be addressed at higher levels. DOE should (1) support a Russian initiative to improve communication not only among Russian site managers but also among ministries and institutes through periodic meetings and perhaps a newsletter, and (2) develop a forum for U.S. site managers to meet regularly and exchange lessons learned.

I. Coordinate MPC&A program activities with activities of related DOE programs to take advantage of opportunities for programs to reinforce one another.

The overlaps in objectives and field activities between different DOE programs are manyfold. They are designed to (1) prevent theft and smuggling of nuclear material; (2) facilitate the downsizing of the Russian nuclear weapons complex and provide alternative employment to its excess scientists and workers; (3) increase transparency in the management of nuclear weapons and materials, particularly in warhead dismantlement and management of excess fissile materials resulting from arms reductions; (4) end production of additional excess fissile material; and (5) reduce the huge stockpiles of excess fissile material as rapidly as practicable. As materials are shipped from site to site pursuant to these programs, the MPC&A aspects obviously should be in the forefront of planning. As MPC&A capabilities of Russian institutes improve, they should be considered as locations for other DOE activities, and as Russian specialists scramble for new income streams, the programs should target the most vulnerable groups.

6. Expand Efforts to Understand the Full Dimension of Critical Long-Term Problems

Whereas the foregoing recommendations are directed to immediate steps, several topics deserve further analysis.

A. Support studies of (1) problems that will confront Russian institutes as they assume full responsibility for sustaining and enhancing MPC&A upgrades, and (2) approaches that can be taken now to help minimize these problems. Particular attention should be given to approaches for generating the funds necessary to sustain the program.

B. Support a study to identify specific cooperative projects that DOE could undertake with GOSATOMNADZOR (GAN), Russia's nuclear regulatory agency, to strengthen GAN's role in MPC&A.

C. Support studies of the capabilities of organized crime groups to penetrate the nuclear establishments of Russia and of the linkages between such groups and organizations in countries of nuclear proliferation concern.

These studies should be joint U.S.–Russian efforts. Also, in view of the centrality of MPC&A concerns to the overall U.S. nonproliferation effort and the changing economic and security conditions in Russia, external reviews of DOE's MPC&A program should be carried out periodically.

EPILOGUE

There are many challenges associated with ensuring the adequacy of material protection, control, and accountability (MPC&A) systems in Russia over the long term. First, the installation of upgraded MPC&A systems that meet international standards at all Russian facilities where direct-use material is located will take large expenditures over a period of a decade or longer. After such systems are installed, substantial annual expenditures for the indefinite future also will be needed to operate them properly. Until such time as the Russian economy recovers, the capacity of the Russian government to pay these expenses will be limited.

Second, there is a need to develop a pool of skilled Russian manpower capable of assuming responsibility for modern MPC&A systems. Although progress has been made, the number of qualified specialists needs to be increased significantly. In addition, the personnel responsible for operating the systems must be committed to avoiding shortcuts or exceptions to prescribed procedures; and they should not hesitate to report to central authorities any irregularities during operation of the systems. Such commitments will depend on professional pride, on the likelihood of severe personal penalties for violations of security requirements, and on an awareness of the importance of their responsibilities both for Russia and for the world.

Third, Russia needs a strengthened industrial and physical infrastructure at the national and local levels for supporting MPC&A systems. This includes indigenous capabilities to produce and service equipment used in the systems, uninterrupted power and communication services, and reliable rail and road networks for transporting material. The degraded condition of the infrastructure is linked directly to economic problems, and it will be difficult to improve it rapidly. Indeed, during the inevitable delay in the revitalization of industrial, transportation, and communication capabilities, special measures will be needed to ensure that the MPC&A systems remain effective.

Fourth, there is a need for a strengthened regulatory framework for ensuring security measures. Although many of the necessary laws and regulations are in place and others are under development, effective enforcement mechanisms to ensure compliance have not yet been established. The concept of an independent regulatory agency (GOSATOMNADZOR) has been stressed repeatedly by Russian officials, but this agency has been plagued by a lack of resources, an unclear mandate, and historical dependence on MINATOM to carry out its responsibilities.

Fifth, there is uncertainty as to the future political leadership of Russia and its commitment to nonproliferation goals. A new government consumed with economic problems could consider MPC&A activities a diversion from the development of programs that promise economic advance. The U.S. government, in cooperation with other governments and international organizations concerned with the possibility of nuclear proliferation, should take steps at the highest political levels to ensure a continuing Russian commitment to nuclear nonproliferation.

Sixth, there is the reality that cooperation could be set back by U.S.– Russian disagreements over nuclear policies, such as the appropriateness of Russian nuclear exports to Iran and other countries of concern. Skill, patience, and perspective will be required to ensure that such disputes do not jeopardize the achievement of enhanced security for both sides.

Finally, there is the challenge on the U.S. side of maintaining the momentum of the program over a possibly extended period. No doubt there will be continuing concerns over the levels of budgetary support for the program, wavering attention by high-level U.S. government officials, waning interest of leading U.S. specialists in traveling to Russia, and consternation with Russian leniency in nuclear export policies. Political disagreements, such as conflicting views on developments in Kosovo, also could disrupt cooperative efforts.

Notwithstanding these formidable challenges, the program of MPC&A cooperation presents an unusual and valuable opportunity to promote the interests of both countries. Vast resources were spent by each side in developing nuclear arsenals, and the amounts being spent in moving away from the nuclear abyss are slight in comparison. Indeed, the expenditure of funds to help ensure the security of nuclear material in Russia is an extraordinarily cost-effective investment in strengthening the national defense of the United States. Seen in this light, the United States should press forward with increased vigor and determination. As noted in the closing of the 1997 report: Seize the opportunity!

BIBLIOGRAPHY

A. Unpublished U.S. Department of Energy Materials

DOE MCP&A Strategy Paper: Consolidating Nuclear Materials in Russia, 15 February 1999, (draft).

Emergency Material Protection, Control, and Accounting Sustainability Measures, January 1999.

Guidelines for Material Protection, Control, and Accounting Upgrades at Russian Facilities, Los Alamos National Laboratory and Sandia National Laboratory, December 4, 1998.

Haase, Michael et al. "MPC&A Upgrades at the Institute of Theoretical and Experimental Physics," August 9, 1998.

Haase, Michael et al. "MPC&A Upgrades at the Moscow State Engineering Physics Institute," August 10, 1998.

MPC&A Consumer Report: Physical Protection Equipment, Sandia National Laboratory, 12 February 1999.

MPC&A Long-Term Viability Program: An Assessment of MPC&A Suppliers Located in the Russian Federation, Brookhaven National Laboratory, February 25, 1999 (Draft).

B. Unpublished U.S.–Russian Agreements and Statements

Agreement Between the Department of Defense of the United States and The Ministry of the Russian Federation for Atomic Energy Concerning Control, Accounting, and Physical Protection of Nuclear Material (2 September 1993 with 20 January 1995 Amendment).

Agreement Between the United States and the Russian Federation Regarding Cooperation to Facilitate the Provision of Assistance (4 April 1992).

Agreement Between the United States and the Russian Federation Concerning the Safe and Secure Transportation, Storage, and Destruction of Weapons and the Prevention of Weapons Proliferation (17 June 1992).

Joint Statement on Cooperation between the Russian Ministry of Defense and the United States Department of Energy on Control, Accounting, and Physical Protection of Nuclear Materials, (June 16, 1996).

Joint U.S.–Russian MPC&A Steering Group, "Unified U.S.-Russian Plan for Cooperation on Nuclear Materials Protection, Control, and Accounting Between Department of Energy Laboratories and the Institutes and Enterprises of the Ministry of Atomic Energy Nuclear Defense Complex," (1 September 1995).

Protocol of the Fourth Meeting of the Joint U.S.–Russian MPC&A Technical Working Group (10 March 1995).

Protocols of the Joint Coordinating Committee of the U.S. Department of Energy and the Federal Nuclear and Radiation Safety Authority of Russia (18-19 October 1995, 17-19 September 1996, 21-19 May 1997, 20-21 November 1997, 8-10 September 1998).

Protocol of the Working Meeting of the U.S. Secretary of Energy and the Commander-in-Chief of the Russian Federation Navy and the Vice-President of the Russian Research Center "Kurchatov Institute" Concerning Cooperation on Nuclear Materials Protection, Control, and Accounting (12 December 1997).

Record of the Meeting of the Third U.S.–Russian Technical Working Group Meeting (6-10 March 1995).

C. Materials Published in Russia
All-Russian Research Institute of Automatics (undated booklet)
AO "Escort Center" (undated booklet)
Eleron: Thirty-Five Years, February 1998 (in Russian)
"Eleron" Special Scientific and Production State Enterprise (undated catalogue of products)
Foreign Trade Organization "Safety Ltd." (undated booklet)
Instruments, vol. 9, 1998 (in Russian)
Joint Institute for Nuclear Research-Dubna, 1980
Moscow State Engineering Physics Institute Master in Nuclear MPC&A, June 1998
Moscow State Engineering Physics Institute Master in Nuclear MPC&A, 1997
Ordinance #264 of the Government of the Russian Federation

D. Other Articles and Published Materials
Albright, D., F. Berkout, and W. Walker. 1997. Pp. 50–59, 94–116 in Plutonium and Highly Enriched Uranium: World Inventories, Capabilities, and Policies. Oxford: Oxford University Press.

Bukharin, O. 1996. Security of fissile materials in Russia. *Annu. Rev. Energy Environ.* 21:467–496.

Bunn, M., and J. Holdren. 1997. Managing military uranium and plutonium in the United States and the former Soviet Union. Annu. Rev. Energy Environ. 22:403–486.

Bunn, M., O. Bukharin, J. Cetina, K. Luongo, and F. von Hippel. 1998. Retooling Russia's Nuclear Cities. Bull. Atomic Scientists, 54 (Sept./Oct.):44-50.

Clinton, W. J. 1999. State of the Union Address. Washington, D.C., January 19.

Deutch, J. 1996. The threat of nuclear diversion. Testimony Before the Permanent Subcommittee on Investigations of the Senate Committee on Government Affairs, March 20.

DOE (U.S. Department of Energy). 1997. Partnerships for Nuclear Security. Washington, D.C.: Office of Arms Control and Nonproliferation.

DOE (U.S. Department of Energy). 1998a. MPC&A Program Strategic Plan. Washington, D.C.: Office of Arms Control and Nonproliferation.

DOE (U.S. Department of Energy). 1998b. Partnerships for Nuclear Security. Washington, D.C.: Office of Arms Control and Nonproliferation.

Doyle, J., and S. Mladineo. 1998. Viewpoint: Assessing the development of a modern safeguards culture in the NIS. Nonproliferation Rev. 5 (Winter):91-100.

Ewell, E. 1998. NIS nuclear smuggling since 1995: A lull in significant cases?" Nonproliferation Rev. 5 (Spring-Summer).

Jenkins, B. 1998. Viewpoint: establishing international standards for physical protection of nuclear material. Nonproliferation Rev., 5 (Spring-Summer)

Kempf, C. R. 1998. Russian Nuclear Material Protection, Control, and Accounting Program: Analysis and prospect. Pp. 40-44 in Partnership for Nuclear Security. Washington, D.C.: Office of Nonproliferation and Arms Control.

Luongo, K., and M. Bunn. 1998. A nuclear crisis in Russia. Boston Globe, December 29.

NRC (National Research Council). 1989. Material Control and Accounting in the Department of Energy's Nuclear Fuel Complex. Washington, D.C.: National Academy Press.

NRC (National Research Council). 1997. Proliferation Concerns: Assessing U.S. Efforts to Help Contain Nuclear and Other Dangerous Materials and Technologies in the Former Soviet Union. Washington, D.C.: National Academy Press.

Perry, T. January 1999. Growing threats to Russian nuclear material security: The imperative of an expanded U.S. response. Draft paper.

Potter, W. 1998. Outlook for the adoption of a safeguards culture in the former Soviet Union," J. Nucl. Mater. Manage. 26 (Winter):22-24.

Richardson, B. 1998. Russia's recession: The nuclear fallout. Washington Post, December 23, A23.

Richardson, B. 1999. Remarks at the 7th Carnegie International Non-Proliferation Conference, Carnegie Endowment for International Peace, Washington, D.C., January 12.

Ryabev, L. 1999. Remarks at the 7th Carnegie International Non-Proliferation
    Conference, Carnegie Endowment for International Peace, Washington,
    D.C., January 11.

APPENDIX
A

Recommendations from the 1997 NRC Report
Proliferation Concerns: Assessing U.S. Efforts to Help Contain Nuclear And Other Dangerous Materials And Technologies in the Former Soviet Union

1. Sustain the Program

• Continue to fund MPC&A efforts in the FSU [Former Soviet Union] at least at the level of fiscal year 1996 for several more years and be prepared to increase funding should particularly important high-impact opportunities arise.

2. Indigenize MPC&A Capabilities

• Continue to emphasize the importance of MPC&A as a nonproliferation imperative at the highest political levels in the FSU.

• Prior to initiating MPC&A projects at specific facilities, obtain assurances at both the ministry and the institute levels that the upgrade programs will be sustained after improvements have been made. Financial incentives, such as support for related research activities, should be considered as a means to stimulate long-term commitments.

• Involve institute personnel to the fullest extent possible in determining how to use available funds for upgrades.

• Give greater emphasis to near-term training of local specialists.

• Reward those institutes that are making good progress in upgrading MPC&A systems by giving them preference for participation in other U.S.-financed cooperative programs.

• Encourage the establishment of new income streams that can provide adequate financial support for MPC&A programs in the long term, such as earmarking for MPC&A programs a portion of the revenues from Russian sales of HEU.

• Rely increasingly on domestically produced and locally available equipment for physical protection, detection, analysis, and related MPC&A tasks.

3. Simplify the Problem

• In Russia, encourage consolidation of direct-use material in fewer buildings, at fewer facilities, and at fewer sites.

4. Minimize the Opportunities in Russia to Bypass MPC&A Systems

• Ensure that all stocks of direct-use material are encompassed in the program, including icebreaker nuclear fuel, supplies at naval facilities, and off-specification and scrap material.

• Encourage rapid development of a comprehensive national material control and accounting system in Russia and the prompt incorporation of all existing direct-use material into that system.

• In Russia, increase support of GAN as an important independent agency by assisting it in developing MPC&A methodologies, training inspectors, obtaining staff support from research institutions, and procuring necessary equipment for MPC&A inspections.

• Encourage a system of incentives, possibly including monetary rewards, that will stimulate participants in MPC&A programs to report promptly to the central authorities any irregularities in the implementation of MPC&A systems.

• Emphasize the importance of developing a culture among MPC&A specialists that does not tolerate shortcuts or exceptions in implementing MPC&A systems.

5. Enhance the Program

• Emphasize MPC&A approaches that respond to threat scenarios that are appropriate for the FSU, recognizing that they may differ from the threat scenarios used in the United States.

• Recognize that in the near term it may be necessary to install systems that fall short of internationally accepted standards in anticipation of subsequent refinements. In this regard, use appropriate MPC&A measures, whether they involve high-tech or low-tech approaches.

• In Russia, give greater attention to MPC&A of direct-use material during transport within and between facilities.

• Promote greater communication and cooperation among ministries and facilities involved in MPC&A in each of the countries where bilateral programs are being implemented.

• In Russia, encourage more active involvement of the Ministry of Interior in the planning, testing, and implementation of physical security systems.

Source: NRC, *Proliferation Concerns*, pp. 11–15.

APPENDIX

B

Terms of Reference

(excerpted from a June 23, 1998 letter
from the National Research Council to Brookhaven National Laboratory)[23]

The assessment will be a follow-on activity to our earlier assessment set forth in 1997 in *Proliferation Concerns: Assessing US Efforts To Help Contain Nuclear and Other Dangerous Materials and Technologies in the Former Soviet Union*, National Research Council, National Academy Press, 1997. Given the importance of MPC&A programs in the former Soviet Union in promoting our national security interests and the extensive DOE activities during the past two years, a new assessment seems appropriate. We understand that this external assessment of DOE activities will complement an internal review being led by Brookhaven National Laboratory.

Our assessment will be directed primarily to programs in Russia. We will revisit the scope of the threat of theft or diversion of unirradiated HEU or separated plutonium, taking into account the many new insights gained by DOE during the past two years. We will consider, for example, the number and vulnerability of Russian sites where HEU and plutonium are located, the amount of material involved, and the effectiveness of MPC&A systems that are in place and under development. Also, using the recommendations in our earlier report as an initial checklist for reviewing activities, we will identify both successes and weaknesses in cooperative approaches to date; and we will extract lessons learned that should be taken into account in future activities.

We will give special attention to progress in "indigenization" within Russia of MPC&A capabilities: (a) the development of a cadre of committed Russian MPC&A specialists who embrace a culture that does not tolerate violations of the principles of sound MPC&A programs; (b) the development of a technical infrastructure that can provide both MPC&A equipment and services to Russian facilities; and (c) the acceptance within the Russian Government and at the facility level of commitments to provide the necessary priority to MPC&A that will

---

[23] The contract with BNL referenced this letter in the Scope of Work.

sustain momentum in the programs over the long term. Also, we will consider how DOE can help ensure that Russian counterparts will continue to improve MPC&A systems after the initial upgrades have been installed at Russian facilities.

In carrying out the project, we will establish a committee of specialists. Initially, they will meet in Washington with DOE officials, American providers of equipment for the program, and other interested parties. Insights from the Brookhaven review will be of considerable benefit to the committee. An early topic for consideration by the committee will be the approaches of DOE in providing support for the many diverse cooperative activities scattered across Russia. We recognize that with the rapid growth of the program there are severe personnel demands both on DOE management and on the American MPC&A specialists leading the effort; an understanding of this reality is important in assessing the approaches that have been adopted and in suggesting future steps. We anticipate that committee members will visit several DOE laboratories to obtain inputs from their specialists.

We are planning a two-week visit by the committee to Russia during the fall of 1998. One week will be spent in the Moscow area. There they will consult with MINATOM officials, return to several sites that our previous committee visited two years ago, visit several sites where DOE has "finished the job," and observe progress in developing the technical infrastructure for supporting MPC&A activities over the long term. During the second week, the committee will visit several locations outside Moscow where MPC&A upgrades are in progress.

Following the visit to Russia, the committee will continue its consultations with specialists from DOE and the laboratories, with a view to completing its report in about seven months. Reports resulting from this effort shall be prepared in sufficient quantity to ensure their distribution to the sponsor and to other relevant parties, in accordance with Academy policy. Reports may be made available to the public without restrictions.

APPENDIX

C

Biographies of Committee Members

Richard A. Meserve (*Chairman*) is a partner in the law firm of Covington and Burling. He holds a law degree from Harvard Law School and a Ph.D. in applied physics from Stanford University. Earlier in his career, he served as clerk for Supreme Court Justice Harry Blackmun and as legal counsel and senior policy analyst in the White House Office of Science and Technology Policy. Dr. Meserve has served as chair or vice-chair of a number of National Research Council committees, including the Board on Energy and Environmental Systems, the Committee on Declassification of Information for the Department of Energy Environmental Remediation and Related Programs, and the Panel on Cooperation with the USSR on Reactor Safety.

John F. Ahearne is currently director of the Sigma Xi Center and adjunct professor at Duke University. He has a Ph.D. in physics from Princeton University and a bachelor's degree in engineering from Cornell. He served as deputy and principal deputy assistant secretary of defense from 1972 to 1977, as deputy assistant secretary of energy from 1977 to 1978, and as commissioner of the U.S. Nuclear Regulatory Commission from 1978 to 1983 (Chairman, 1979–1981). Dr. Ahearne was also vice-president and senior fellow of Resources for the Future. Prior to his current position, he served as Executive Director of Sigma Xi. He is a member of the National Academy of Engineering and has served on many National Research Council committees, including committees on plutonium disposition and risk management. He is vice-chair of the Council's Board on Radioactive Waste Management.

Don Jeffrey (Jeff) Bostock recently retired from Lockheed Martin Energy Systems, Inc., as vice-president for Engineering and Construction with responsibility for all engineering activities within the Oak Ridge nuclear complex. Prior to assuming that position, he served as vice-president of Defense and Manufacturing and manager of the Oak Ridge Y-12 plant, a nuclear weapons fabrication and manufacturing facility. His career at Y-12 included engineering and managerial positions in all of the various manufacturing, assembly, security, and program management organizations. He also served as manager of the Paducah Gaseous Diffusion Plant providing uranium enrichment services. Mr. Bostock has a B.S. in industrial engineering from Pennsylvania State University and an M.S. in industrial management from the University of Tennessee. He is a graduate of the Pittsburgh Management Program for Executives.

William C. Potter is a professor and director of the Center for Nonproliferation Studies at the Monterey Institute of International Studies (MIIS). He also directs the MIIS Center for Russian and Eurasian Studies. He is the author or editor of 12 books, including *Dismantling the Cold War: U.S. and NIS Perspectives on the Nunn-Lugar Cooperative Threat Reduction Program* (1997). He has served as a consultant to the U.S. Arms Control and Disarmament Agency, Lawrence Livermore National Laboratory, RAND Corporation, and Jet Propulsion Laboratory. His present research focuses on nuclear exports, nuclear safety, and proliferation problems involving the post-Soviet states. He is a member of the Council on Foreign Relations and the International Institute for Strategic Studies and serves on the International Advisory Board of the Center for Policy Studies in Russia and the International Institute for Policy Studies in Belarus. Dr. Potter was an adviser on the Kyrgyzstan delegation to the 1995 Nonproliferation Treaty Review and Extension Conference.

APPENDIX

D

Site Visits and Meetings of the Committee

RUSSIA

Visits to sites where MPC&A upgrades are being installed:

Institute of Physics and Power Engineering (Obninsk)
Institute of Theoretical and Experimental Physics (Moscow)
Joint Institute of Nuclear Research (Dubna)
Kurchatov Institute of Atomic Energy (Moscow)
Luch Scientific Production Association (Podolsk)
Moscow Engineering Physics Institute (Moscow)
Scientific Research Institute of Atomic Reactors (Dmitrovgrad)

Visits to oganizations that produce MPC&A equipment:

All Russian Research Institute of Automatics (Moscow)
Eleron (Moscow)
Escort Center (Moscow)

Visits to organizations that manage MPC&A training programs:

Institute of Physics and Power Engineering (Obninsk)
Kurchatov Institute of Atomic Energy (Moscow)
Moscow Engineering Physics Institute (Moscow)

Meetings with government agencies and regulatory bodies:

GOSATOMNADZOR (Moscow)
GOSATOMNADZOR (Dmitrovgrad)
Ministry of Atomic Energy (Moscow)
Ministry of Finance (Moscow)

Meetings at other organizations:

International Science and Technology Center (Moscow)

UNITED STATES

Meetings on the overall MPC&A Program:

Brookhaven National Laboratory
Department of Energy
Russian–American Nuclear Security Advisory Council

Meetings on implementation of the MPC&A Program:

Lawrence Livermore National Laboratory
Los Alamos National Laboratory
Oak Ridge National Laboratory
Sandia National Laboratory

APPENDIX

E

MPC&A Program Guidelines

Mission

The mission of the MPC&A program is to reduce the threat of nuclear proliferation and nuclear terrorism by rapidly improving the security of all weapons-usable nuclear material in forms other than nuclear weapons in Russia, the NIS [Newly Independent States], and the Baltics.

Goals and Strategies

1. **Reach Agreement for MPC&A Cooperation with all Sites in Russia, the NIS, and the Baltics Containing Weapons-Usable Nuclear Material in Forms Other than Nuclear Weapons:**
   A. Overcome mutual cold war suspicions, lack of technical working relationships, security issues at closed nuclear cities, and language and cultural differences.
   B. Establish contracts or other agreements to upgrade MPC&A at all facilities within these sites, which store, process, or transport Pu of HEU.

2. **Implement Systematic and Rapid MPC&A Upgrades at all Sites:**
   A. Concentrate MPC&A efforts on the most attractive materials for nuclear weapons, namely, HEU (20% and greater) and Pu (excluding Pu in irradiated fuel).
   B. Install comprehensive, technology-based MPC&A systems that are consistent with international standards, such as IAEA INRFCIRC/225 and the IAEA Guidelines for State Systems for Accounting and Control (SSAC), which are appropriate for the unique conditions at each site and effective for securing nuclear material against insider and outsider threats.

C. Use proven MPC&A methods and technologies.

D. Use both indigenous (Russian, the NIS, and the Baltics) and foreign technologies, depending on the technical merits. Indigenous technologies, when available, may have advantages in terms of cost, maintainability, and acceptance, and other factors. Foreign technologies, on the other hand, may have advantages in terms of uniqueness, availability, reliability, track record, and other factors. Decisions on using these technologies is to be made jointly, taking all relevant factors into account.

E. Transfer full responsibility for the long-term operations of upgraded MPC&A systems to our partners after the completion of cooperative upgrades and provisions of associated manufacturer guarantees.

F. Assist guard forces with radiocommunications, investigative techniques, and other mechanisms/capabilities to improve guard force operations, without providing training in the use of force or purchasing weapons.

3. **Ensure Long-Term Effectiveness of Improved MPC&A Systems:**

A. Establish MPC&A training programs.

B. Strengthen national nuclear regulatory systems and national standards for MPC&A.

C. Foster indigenous production and maintenance of MPC&A equipment.

D. Conduct annual reviews of vulnerabilities and hardware to determine if additional MPC&A upgrades are required to meet changing conditions.

4. **Achieve Technical Integrity and Openness**

A. Carefully protect sensitive information and technologies in all facets of the program.

B. Sustain MPC&A program as a multilaboratory program operating under DOE guidance and oversight. Ensure U.S. experts (DOE, laboratory, and contractor personnel) work together as a unified team committed to common objective.

C. Ensure that the basic operating principle of the MPC&A program aligns with capabilities and responsibilities. Assign work according to demonstrated capability and capacity in accomplishing program objectives.

D. Follow a disciplined approach in planning and executing projects. Assess proposed work in terms of is it needed; is it timely; is it cost effective; have all unnecessary activities and costs been eliminated?

Frequently Used MPC&A Upgrades

1. Physical protection systems: locks, fences, barriers, gates, badging systems, and interior and exterior sensors, including video cameras and motion detectors.
2. Alarm systems and computers to process data from sensors, such as closed-circuit television and communication systems to improve response to alarms.
3. Nuclear material detectors installed at pedestrian and vehicle portals, which detect attempts to remove nuclear material, including hand-held detectors for random guard-force checking.
4. Tamper-indicating devices to prevent unauthorized removal, computerized MPC&A systems, including barcode systems, to track nuclear material inventory.
5. Perimeter clearing and structural improvements to improve physical protection.
6. Computerized material accounting systems to maintain physical inventory and non-destructive assay measurements.

Source: DOE, *MPC&A Program Strategic Plan*, pp. 8–9.